William Judkins Conklin

The Relations of Epilepsy to Insanity and Jurisprudence

William Judkins Conklin

The Relations of Epilepsy to Insanity and Jurisprudence

ISBN/EAN: 9783337319335

Printed in Europe, USA, Canada, Australia, Japan

Cover: Foto ©berggeist007 / pixelio.de

More available books at **www.hansebooks.com**

THE RELATIONS

OF

EPILEPSY TO INSANITY AND JURISPRUDENCE.

BY W. J. CONKLIN, M. D.,

Assistant Physician of Southern Ohio Lunatic Asylum,
Dayton, Ohio.

Read before the Ohio State Medical Society, April 6th, 1871.

The Relations of Epilepsy to Insanity and Jurisprudence.

BY W. J. CONKLIN, M. D.,

Assistant Physician of Southern Ohio Lunatic Asylum, Dayton, O.

Epilepsy and insanity may sustain various relations to each other.

1. The two diseases may originate at the same time and be due to a common cause. The following cases are instances of this relation :

L. L., æt. sixteen, being alone one night during the absence of the family in which she served, was terribly frightened by two men endeavoring to enter her sleeping apartment. She was thrown into convulsions and became delirious, when they ceased. Her mind rapidly failed. The fits recurred at short intervals, usually at night. She is now, ten years after the occurrence, a hopeless dement, and apparently from loss of muscular co-ordination is barely able to pronounce her own name.

F. McK., æt. 27 ; maternal cousin insane ; for three years he had given himself up to all manner of excesses and dissipations. He experienced a regular epileptic seizure, from which time his friends noticed symptoms of mental derangement. The convulsions returned at intervals of four weeks ; after the third fit he became violently maniacal and was admitted into the asylum.

2. Epilepsy may be developed in the course of chronic insanity. Beyond question, the degenerated condition of the nervous system which the insanity itself indicates is a fertile soil for the springing up of nervous disorders of all kinds ; but the frequency with which epilepsy occurs as a sequence among the chronic insane, by no means an inconsiderable per cent of our population, is not great.

The tendency of many organic brain diseases is to epileptiform seizures. A large number of the cases of convulsions occurring in the old residents of an asylum are due to the presence of a tumor, exostosis, softening or other cerebral lesions, which we are, oftentimes, able to accurately diagnose during life. These cases should be excluded from epilepsy proper. From our limited experience, we are inclined to consider true epilepsy as an infrequent result of insanity. Many so called cases of the relation under consideration are really examples of masked epilepsy : others may be cases of undiscovered epileptic vertigo, which have passed into the graver form of the disease.

It may be considered a fact, that the appearance of convulsions in an attack of insanity is always of grave portent, and is usually the forerunner of death. Within the past two years, seven cases have occurred in this institution confirmatory of this proposition. These include all cases in which the convulsions appeared primarily in the course of the mental disturbance, and they all died in from twenty-four hours to nine days after the supervention of the convulsions. It may be well to remark, that none of them presented the peculiar character of epileptic insanity.

In two of the above cases the insanity had a duration of one year ; in three cases it was of three years duration ; in one it dated back ten years, and in another twenty years.

Post mortem examinations of the brain were made in two cases. In one of them, in whom the mental symptoms were of three years duration, the dura mater was found attached to the skull, and, in places, along the longitudinal fissure, was firmly united to the arachnoid. The vessels of the arachnoid and pia mater were enlarged and engorged with blood. About the middle of the right anterior lobe there was a loss of brain substance, measuring one half of an inch in length and one quarter of an inch in breadth. The puncta vasculosa of the white substance were quite numerous and distinct. The corpus striatum and optic thalamus of the right side were much less firm in texture than the corresponding parts of the opposite side.

The autopsy of the second case in which the insanity was of twenty years standing, revealed nothing abnormal, save an unusual adherence of the dura mater to the calvarium and considerable congestion of the left hemisphere. The brain substance was not microscopically examined in either case.

3. Another relation, important in a medico-legal as well as psychical point of view, is that of transformation or replacement, in which the mental manifestations take the place of the regular convulsion. The morbid action in the sensory centres seems at one time to be transferred wholly or in part to the motory centres, giving rise to muscular convulsions; and at another time to the mind centres, giving rise to convulsive ideas. Epileptics often speak of the torture and anguish they feel, when they have gone longer than usual without the fit, and the feeling of relief experienced after it is over.

It is not unusual in the history of the insane to find some previous bodily lesion becoming latent or masked on the accession of the mental derangement, even though the latter may be the result of the former. This occurs frequently among the phthisical, when on the supervention of insanity the tuberculization is for a time arrested. Dr. Burrows tells of an eloquent divine, who was always maniacal when free from pains in the region of the spine, and sane when the pains returned to that spot. We know that physical lesions will often for the time stop epileptic seizures.

K. F., age 22, subject to epilepsy from her childhood, fell into the fireplace in a fit, and was terribly burned over the chest, abdomen and thighs. Her fits, which had been previously severe, and rarely missing more than two days at a time, ceased, and did not return for twelve months, during which the wounds were healing. As soon as the burn healed, the fits returned with their former frequency and severity.

The arrest of the fits from any cause often produces an increase in the severity of the mental symptoms. This is seen in the effect of medical treatment. Often when by the exhibition of some medicine the fits are lessened in frequency, the patient is rendered more irritable and peevish, it may be violent. In a discussion at the last meeting of Asylum Superintendents,* Dr. Gundry said, "One Doctor gave medicine, and six months afterwards the cry was raised, 'You cured him of epilepsy, and the people wish he had fits again, because he is so infernal cross.' I have found in many cases, I may say, when the fits were intercepted, the same nervous excitement takes their place."

Dr. Walker, of the Boston Lunatic Asylum, said, "There are two or three cases, as stated by Dr. Earle, when the epileptic fits have been decreased, and they have been made irritable and are more violent. They were not subject to it when the fits came regularly, and it was thought proper to let them have a fit occasionally to sweeten them."

Every asylum presents cases in which so long as the convulsions occur with their usual frequency, no unusual mental symptoms present themselves, but as soon as the seizures cease, the most marked mental disturbance ensues, often violent in nature, and attended with delusions and hallucinations; with the reappearance of the convulsions the

---

* American Journal of Insanity, October, 1870.

patients resume their normal mental state. Still more frequently occur those cases, in which brief but violent outbursts of fury—blind impulses to destroy, overpower the mind and impel to homicidal, suicidal or destructive acts—occasionally replace special convulsive seizures. The delusion or hallucination which often becomes apparent just before an outbreak may be justly compared to the aura; "the moral instability to the irregular involuntary character of the muscular action; the sudden impulse with the fit, and the temporary unconsciousness with the coma."

Many instances of replacement will appear in the progress of this paper. In this connection we will relate but one case, taken from Dr. Schupman. An epileptic, during convalescence from a severe attack of cholera, was free from his usual epileptic seizures. He became restless, agitated, conceived that he felt some living thing moving in his stomach, and finally declared that he was pregnant. The slightest contradiction led to such violence that it become necessary to confine him to a chair. Many expedients were adopted to remove his delusion, among them an accouchment was imitated. He retained his delusion in spite of treatment, until the attacks of epilepsy returned, when it immediately disappeared.

4. Epilepsy may stand as the cause of insanity. This relation has a two-fold nature: It may be a predisposing cause, or it may be an exciting cause.

Epilepsy is now generally recognized as a strongly marked hereditary disease. Statistics show that epileptic children are more likely to be born from an insane than from an epileptic parentage. It may be stated as a general law, that the same neurosis from which the parent suffers is not necessarily transmitted to the offspring, but save in suicidal insanity alone, the type or character of the disease is more liable than otherwise to be transformed. Families of epileptic parents sometimes exhibit insanity in one of its members, epilepsy in a second, peculiarities of character in a third, and chorea or some other neurosis in another member. Rarely do we find individuals tainted with epilepsy to be entirely healthy. They nearly all possess some vice of constitution which renders them incompetent to meet successfully the ordinary trials and troubles of life; often they exhibit the insane or epileptic neurosis, which are very similar to each other, and render the individual more susceptible to the ordinary exciting causes of insanity. Dr. A. Foville* gives an interesting account of seven observations upon epileptic families. He traced two families for three generations, and four families through four generations. I append his table, which comprises 129 persons, either ancestors or descendents of the special epileptic patient, who is designated by the initial of his name.

* Annales Medico-Psych., Vol. 11., p. 229.

SYNOPSIS OF SEVEN OBSERVATIONS OF THE FAMILIES OF EPILEPTICS.

OBSER-VATION.	FIRST GENERATION.	SECOND GENERATION.	THIRD GENERATION.	FOURTH GENERATION.
I.	1 Father, epileptic........	8 Sons, epileptic...........	3 Grandsons, epileptic.........	1 Great-Grandson, epileptic.
II.	1 Grandfather, insane, epileptic.........	1 Father, insane, epileptic	{ 1 G. Male, epileptic............ 2 Sisters, insane. .... 1 Brother, intelligent but odd. }	?
III.	1 Grandfather, insane...	{ 1 Father, insane, married. 1 Mother, epileptic........ }	{ 1 Sister, presumed sane........ 1 M. Female, epileptic .... }	{ 3 Children, died in infancy. 3 Presumed sane. 1 Daughter, hysterie. 1 Daughter, epileptic. }
IV.	1 Father, epileptic..	{ 1 Maz. Female, epileptic, married .... }	{ 9 Children died early,........ 1 Son, presumed sano........ 2 Sisters, died early. 4 Brothers, died early. }	?
V.	1 Grandmother, insane..	1 Mother, presumed sano,	{ Bac. epileptic......... 1 Sister, scrofulous. }	4 Children, died early.
VI.	1 Father, epileptic. ..... , 1 Uncle, imbecile	{ 1 Sister, died young. 1 N. insane, epileptic...... 1 Brother, epileptic..... 1 Brother, insane.. ..... 4 Others presumed sanc 1 Cousin, insane......... 2 Cousins, presumed sane }	{ 1 Daughter, epileptic........ 1 Daughter, presumed sane...... 3 Nephews, epileptic. 27 Nephews and Neices, sane. ? .... }	{ 2 Little Children, so far sane. 2 Little Children, so far sane. Married, but no children. }
VII.	1 Grandfather, epileptic,	{ 1 Father, epileptic......... 2 Aunts, sane......... .... }	{ 1 It. epileptic......... 5 Brothers and Sisters, died early 1 Daughter, scrofulous....... 2 Daughters, presumed sane..... 4 Cousins, sane .... }	{ Not married. 5 Nephews and Neices, sane. No children. }
Total.	8	27	72	22

Casting this table in a different form we find :

Generation.	Epileptics.	Insane.	Presumed Sane.	Children Died early.	Total.
1	5	3			8
2	14	3	10		27
3	11	2	39	20	72
4	2	1	12	7	22
Total, . .	32	9	61	27	129

It is but fair to presume that some of the sixty-one *presumed sane* at the time the observations were made, would, if their subsequent history were traced, increase the number of the epileptic and insane.

M. Herpin, in 380 relatives of 68 epileptics, found ten epileptic, twenty-four insane, and forty-four others laboring under some nervous disorder. He calculates that the relatives of epileptics exhibit epilepsy about five times more frequently, and insanity about twenty four times more frequently than does the French population generally.

The forty-two cases of insanity due to epilepsy, admitted into this institution, in which the parties were only interrogated in regard to the prevalence of mental disorder in the ancestry, we find :

	M. E. T.
No Hereditary Predisposition......	13, 8, 21
Hereditary Predisposition....	8, 6, 14
Unknown..............	4, 3, 7
Total.................	25, 17, 42

The frequency with which epilepsy occurs as an exciting cause of insanity can not be exactly determined. The testimony of nearly all observers agree as to the ultimate result upon the intellectual faculties. Esquirol* found among 339 epileptic females in the Saltpetre, 269 in a state of mental alienation, and this proportion, though large, would doubtless be increased if the final history of the remaining ones

---

* Maladies Mentales, t. 1., p. 142.

were traced. Trousseau* writes: as the fits recur, increase in frequency, in proportion as the disease progresses, the faculties fail, are impaired, become gradually extinct, and insanity follows.

Dr. Echeverria† holds the following language: "Epilepsy contrasts singularly with the other diseases in the deep marks it impresses on the organic and moral constitution of the individual. Scarcely do we find confirmed epileptics, who, on careful examination, would not afford evidence of this statement. The intellectual impairment originates by degrees, or by imperceptible stages—nay, it may be tardy in its display, though it affects the instincts and emotions from the beginning."

Dr. Sievcking‡ remarks: "A single fit may occur never to return, and without leaving any trace of disease. More commonly, however, unless the disease be arrested and the habit broken, the fits recur with gradually increasing frequency, and it is then that we soon discover the intellectual faculties begin to fail."

Dr. Blandford§ writes: "This is a condition quite distinct from the loss of memory and general dementia, which, if the fits are at all frequent, gradually encroaches upon the mind, and almost without exception terminates the career of every epileptic, even when there has been no insanity so called throughout the entire illness."

Dr. Reynolds,‖ however, analyzed sixty-two cases in reference to their mental condition, and arrived at conclusions differing from the authorities above cited. He concludes:

1. That epilepsy does not necessarily involve any mental change.

2. That considerable intellectual impairment exists in some cases, but that it is the exception and not the rule.

---

* Clinical Lectures, p. 67.
† Echeverria on Epilepsy, p. 302.
‡ On Epilepsy, p. 59.
§ Insanity and its Treatment, p. 70.
‖ Epilepsy, p. 46.

3. That women suffer more severely and more frequently than men.

4. That the commonest failures is loss of memory, and this if regarded in all degrees is more frequent than integrity of that faculty.

5. That apprehension is more often found preserved than injured.

6. That ulterior mental changes are rare.

7. That depression of spirits and timidity are common in the male sex, but not in the female. That excitability of temper is found in both.

He found the class in which there was no discoverable departure from normal mental action to be thirty-eight per cent. of the whole number. This result, so different from that reached by most observers, should receive a special investigation.

Dr. Reynolds remarks, a page or two in advance, " From the following analysis I have excluded all cases of positive insanity, of general organic disease, of distinct cerebral disease, where, although there may have been convulsions, there was not prominently a case of epilepsy, and all instances of simple eccentric convulsions."

That all in the above enumeration, save *positive insanity*, should be excluded, is right and proper; but it hardly seems fair to throw out insanity resulting from epilepsy, in an inquiry which has for its object the ascertaining of the effects of epilepsy upon the mental powers. It must certainly lead to erroneous conclusions, when in analyzing a certain number of cases to ascertain the frequency of a lesion all those cases in which the lesion is best marked are excluded. Epilepsy is a chronic and progressive disease, and the only way of accurately determining its effects upon the organism is by ascertaining the mental and physical condition in the later years of life. Because no mental or emotional irregularity was discoverable at the time his observations were made, docs not prove that such irregularity did not manifest itself in the subsequent history of the case.

Dr. Reynolds also selects loss of memory and apprehension as points of departure in estimating the mental health of the individual. These, however, are certainly qualities of mind in which slight deviations are scarcely discernable, and the failure of which in any notable degree marks advanced disease. The departure from health would probably manifest itself primarily in the affective faculties, irrregularities of which nearly always precede derangement of the intellectual faculties. Unfortunately, on account of limited hospital capacity, nearly all of our asylums exclude insanity complicated with epilepsy because of its incurability, and hence the data for computing what per cent. of the insane is due to epilepsy, are very imperfect.

Dr. Dunglison's statistics of 11,269 cases, compiled from the reports of American asylums, the large majority of which exclude epileptic cases, gives epilepsy as the. eleventh cause in frequency. Taking the annual reports for 1870 of Longview Asylum, State Lunatic Asylum, Pa.; Western Pennsylvania Hospital, and Missouri State Lunatic Asylum, in which institutions the chronic insane are not, so so speak, legislated against, I compile the following table. It must, however, be remembered that the chronic insane are practically excluded, owing to insufficient hospital accommodation and to the greater pressure rightfully made in favor of a recent and probably a curable case.

ASYLUMS.	Whole number of admissions with known cause.	Relative frequency with which Epilepsy occurs as a cause.	Number of males admitted with known cause.	Relative frequency of Epilepsy as a cause among males.	Number of females admitted with known cause.	Relative frequency of Epilepsy as a cause among females.
Longview Asylum................................	1,801	6th	939	3d	862	10th
State Lunatic Asylum, Pa...............	1,560	4th	799	3d	761	6th
Western Pennsylvania Hospital.....	1,602	6th	928	5th	674	6th
Missouri State Asylum	1,120	2d	640	2d	480	5th
Total	6,083	....	3,306	....	2,777	....

Putting this result in a different shape, we find that in 6,083 admissions attributable to known causes, epilepsy is responsible for 6.39. Of 3,306 males, 8.62. Of 2,777 females, 3.74 are caused by epilepsy.

Dr. Nichols, Superintendent of the Government Hospital for Insane, Washington, D. C., in a private letter, after remarking that it is impossible to state with accuracy the per cent. of admissions caused by epilepsy, says, "Nine per cent. of the admissions to this hospital are epileptic." Dr. Kellogg, Assistant Physician New York City Lunatic Asylum, writes me that of 436 males in the asylum, 7 per cent. are epileptic; of 864 females, 4.50 per cent. are epileptic. A census of the weak minded in Scotland, taken several years ago, showed nearly eleven per cent of the whole number to be epileptic. The conclusions warranted by the above data are only approximatively correct. The error is, however, not in favor of epilepsy.

All statistics as to the causation of insanity are to be received with allowance; the above with more than the usual allowance, since they only concern the few insane epileptics who have received admission into an asylum. The proportion which they bear to the whole number, may be judged when we consider the crowded condition of our county and city infirmaries.

From the foregoing we may consider:

1. That epilepsy is a potent cause of insanity.

2. That a larger percentage of male than of female admissions are due to epilepsy.

A question naturally arises in this connection, possessing much interest in a medico-legal sense, as to what is the true mental condition of epileptics. There is no denying the fact that persons subject to periodical attacks of epilepsy are of unsound nervous constitution. The fact that there is an utter unconsciousness at the time of, and for a variable period succeeding the fit, during which coma may ensue, or automatic acts, from the mildest to the most destructive, may be performed ; during which the person is, so to speak, morally dead, shows a vast difference between the class of people under discussion and perfectly healthy ones. True, we are all periodically unconscious during sleep, and may perform acts of which there is afterwards no remembrance, but there is this prominent difference : the latter is a physiological phenomenon, and to a slight extent is under the control of the will; the former is a pathological phenomenon wholly removed from the control of the will, and dependent upon a disease, the tendency of which is to a gradual extinction of the mental faculties.

Esquirol says, "They (epileptics) have exalted ideas, they are very susceptible, irascible, obstinate, difficult to please, odd—they all possess some peculiarity of character."

M. Calmeil writes: "That all epileptics not yet insane are irascible, very impassionable, and disposed to false interpretations."

Baillarger* says: "All authors are agreed in admitting the fact that epilepsy, before leading to complete insanity, produces very important modifications in the intellectual and moral condition of certain patients."

Dr. Burgess† writes: "Epilepsy and chorea are so closely associated with madness as to obtain a special and family character with it."

Dr. Blandford‡ while holding that epilepsy is not the equivalent of insanity, admits that there is good grounds for considering the mental condition of epileptics unsound.

Dr. Browne,§ Ex-Commissioner of Lunacy for Scotland, says: "However much the minds of epileptics may in some respects resemble those of healthy individuals, they differ from all those and differ in the same respect."

---

* Medical Critic and Psych. Journal, Vol. 1.
† Relations of Madness, p. 45.
‡ Op. cit., p. 172.
§ Journal of Mental Science, Oct., 1865.

That persons prominent in the world's history, and others who have achieved eminence in particular callings, have suffered from epileptic seizures does not negative the proposition. Let us, however, briefly glance at the roll. Napoleon is usually quoted as an example of this class. His character certainly has many points in common with the epileptic character, and presents numerous evidences of a diseased nervous system, but I am unable to find authority for the assertion that he was an epileptic. The translator of Trousseau refers to Sir Thomas Watson,* but he distinctly says: "Napoleon is said, I know not upon what authority, however, to have suffered something like epilepsy during sexual intercourse."

Mahomet is also included among the number. The researches of Dr. Weil and Washington Irving confirm the suspicion that he was subject to epilepsy; although his biography is so enshrouded by Arabian fables and traditions, what is known certainly presents evidence strong enough to convince the most sceptical that he was the possessor of an unhealthy nervous organization. Ayesha, one of his wives, and Zeid a disciple, bear witness that he was at one time siezed with violent convulsions, after which he would lie with his eyes closed, foaming at the mouth, and bellowing like a young camel. Cadijah, his first wife, also noticed similar spells while in Mecca, before the pretended revelation of the Koran. Irving† finds the explanation of his conduct in his "enthusiastic and visionary spirit gradually wrought up by solitude, fasting, prayer and meditation, and irritated by bodily disease (epilepsy) into a state of temporary delirium, in which he fancies he receives a revelation from heaven, and is declared a prophet of the most high."

As to Cæsar, we are told that at one time when the Senate was offering him special honor, he acted discourteously

---

* Principles and Practice of Physic, p. 414.
† Life of Mahomet, pp. 61, 335: Vol. 1.

to the Senators, and observing their displeasure, bared his
throat for the knife to the very men assembled to show him
respect. He then excused his conduct on account of the
malady from which he suffered, saying: That those attainted
are unable to speak when standing in public, that they experience shocks throughout their frame, that they suffer
from vertigo, and finally lose consciousness altogether.*

Peter the Great, who was subject to some nervous disorder, considered by many to have been epilepsy, has also
been included in the list.

At the death of his son by Catherine, he was seized with
convulsions; he threw himself prostrate upon the ground
and remained there unattended for three days and nights.

Charity would certainly try to shield his many shortcom-
ings under the garb of disease. The civilization of the
period in which he lived can hardly be responsible for the
neglect of all the civilities and politenesses due to equals,
especially when he is their guest, any more than his, at times,
almost barbarous treatment of the Czarina. It is related,
that when stopping at Magdeburg, on his road to Berlin,
he gave an audience to the select Magdeburg public. During the audience, the Duke of Mecklenburg with his duch-
ess, who was a niece of the Czar, arrived. At the sight of
the latter " Peter started up, satyr-like, clasping her in his
arms and snatching her into an inner room, with the door
left ajar, and there — it is too Samoedic for human speech,
and would excel belief were the testimony not so strong."†
Shortly after, dining with Friedrich Wilhelm and the Queen,
at Berlin, the Czar placed himself by the side of the Queen.
During the meal he experienced a seizure, became convulsed,
gesticulated wildly, danced around the Queen with a knife
in hand, and finally grasped her hand with such violence as
to cause her to shriek out.

---

* Psychologie Morbide; Moreau.   P. 551.
† Frederick the Great; Carlisle.   Vol. 1., pp. 348—350.

Petrarch, Italy's greatest lyric poet, was an epileptic; and see him, all his life long a victim to his passion for Laura, of whose virtues and memory he so delighted to sing, and from whom neither fame nor friends nor time could wean him. He fell in love with her on seeing her in church, and even went so far as to declare his love, though she was already the wife of another. For ten years he was constantly agitated between his love and his reason. Depressed in spirits, he left Avignon and travelled, to rid himself of his passion, but in vain. His love for her was not exclusive, for we learn that the mortification of being the father of an illegitimate son, though he afterward· had an illegitimate daughter by the same woman, he retired to Vancluse, where "his ears were disturbed only by the sounds of nature, and only one female came, a swarthy old woman, dry and parched as the Libyan desert."

From Dr. Browne* I quote the following:

"Conceive the golden-mouthed apostle Bossuet, a victim to the petit mal, terrified by the prospect of lythotomy, losing language, shorn of the glorious gifts which even now give him a prominent place among the orators and defenders of the Christian faith, and think of him as haunted, persecuted, tyrannized over by an ever-recurring ode of Horace, which excluded every other thought and feeling.

"Moliere has been added to this group. Suffice it to say, that he was incapacitated for thought or work, for that was his work, for fifteen days after every fit, that he lived on milk; drew his own portrait in writing " Le Malade Imaginaire;" had a fit while acting the part, which, with consummate skill, he concealed under a laugh; that he was estranged from his wife; chose an old woman critic of his plays; and was denied Christian burial, as much on the ground of his eccentricities as his infidelity.

"Newton's glimpses of ' cycle in epicycle rolled,' ended in

---

* Jour. Ment. Science. October, 1865.

epilepsy and dementia. The tic nerveu and perhaps the opiomania of Madame De Stael ended in delirium. The delicately strung system of Pascal is a sort of lay figure upon which to study the most rare and mysterious neuroses. He lived under incessant attacks of petit mal, and died convulsed. He wrote the bitterest satire and the most generous and genial dissertation on ethics. He was a mathematician, moralist, philosopher, but he believed in charms and amulets, and at all times, however occupied, there was an ever-yawning gulf beside him — a gulf into which he could not divest himself of the apprehension that he might be precipitated."

Hence, we see that these men, great as they were and who are usually considered as examples of epileptics suffering no departure from mental health, when closely scrutinized give unmistakable evidence of a diseased mental organization. These cases, however, constitute the exception; all that could be proven by them would be that such men had intellectual stamina enough to withstand to any degree the injurious effects of the disease. Had their mental power been no greater than the large majority of epileptics, they would have exhibited the failure of the class. Just how far the disease may have affected their private and social relations it is impossible to judge; doubtless much of the irritability which is excused on account of their peculiar surroundings is in reality the outcropping of the disease.

Again, great as these men were, the characters of most of them present some grave moral obliquities, and who will say that some blots which now blacken their fame would have occurred had their nervous systems been healthy.

The mental condition of epileptics may manifest itself prominently in three directions:

1st. In temporary outbursts of delirium or uncontrollable fury.

2d. In a perversion of the moral qualities.

3d. In an attack of regular maniacal excitement, lasting from a few hours to several weeks.

These divisions are to a certain extent arbitrary, it is rare to find them distinct for any length of time; they usually blend together, or occur at different times in the same individual. Still for description it serves our purpose, inasmuch as the first or second condition may for quite a while be the only psychical manifestations of the disease.

We may state in the outset, that epileptic insanity in any of its manifestations is paroxysmal, like the seizures to which it always bears some relation. As the convulsion is the grand de appui, the mental symptoms naturally arrange themselves into, 1st, Those immediately preceding or succeeding the fit; 2d, Those occurring between the fits.

Our first division, characterized by paroxysms of blind fury, and which not infrequently constitute the only departure from normal mental action, possesses great legal interest. In the intervals, which may continue for months or even years, the person manages his business affairs and performs the ordinary duties of life, seemingly with as much tact and judgment as ever; still, close scrutiny reveals a certain mental condition peculiar to the class, prominent in which may be mentioned a changeableness of thought, action, and feeling. He becomes irritable, is greatly annoyed by trivial matters, grows suspicious of those around him, at times is despondent and given to gloomy thoughts, or again he is over-exhilarated and abounds in magnificent projects for the future. As the disease progresses, he becomes at one time quarrelsome, and again, is imbued with religious sentiments. He becomes forgetful, suffers from confusion of thought and loss of apprehension. He leaves off his industrious habits, becomes niggardly, or, it may be, prodigal of his worldly goods, until finally his insanity becomes unquestionable.

This storm of passion—this furor epilepticus—at times

sweeps like a whirlwind over the mind and impels to the most wanton acts. The will of the sufferer seems overwhelmed by the disease, and he is transformed into a perfect automaton, dealing injury and destruction to persons and things around him. The paroxysm may expend its force in harmless action, but usually culminates in some act of violence. The painful nature of the delusions and hallucinations of epileptics may afford a partial explanation of the violence of their actions. Hallucinations of sight, and especially of blood and red colors, occur most frequently. Often just preceding and during the explosion they see armed men, ghosts, assassins, who rush upon them to kill them.

These paroxysms may precede, follow or replace the regular epileptic convulsion. In the majority of cases the mental excitement precedes the convulsion. It is not unusual to hear an attendant say, "Doctor, I don't believe she would have her fits if she did not get so angry. She gets so mad at some one that she throws herself into a fit." In point of fact, the anger is as much beyond the patient's control as is the convulsion which it foreshadows. The fact that the psychical phenomena do replace the seizure is admitted by all; and in some cases, constituting the epilepsia larvalis of Morel, these mental symptoms may be the only manifestations of disease.

The explanation may be that the morbid activity which in uncomplicated epilepsy is restricted to the medulla oblongata, is by some means extended to the cerebrum, causing an epilepsy or convulsion of the brain; for it must be admitted, that the delirium in the suddenness of onset, the violence of its course, the shortness of its duration, and in the resemblance of successive attacks, presents strong analogies to the physical phenomena. However, with our present imperfect knowledge of the pathology of epilepsy, we are unable to wholly explain the above phenomena, or to understand why an attack of furor or mania should occur in connection with the convulsion in one person and not in another, or in the same person at different times. Dr. Blandford* says that "we must suppose that the disturbance in the brain circulation implied by the epileptic seizures does not at once subside, as it does in a patient who in an hour appears to be in all respects sane and unchanged. The reaction, if we may so term it, after the fit or fits brings about a great disturbance of the circulation, and, according to the degree thereof, symptoms of insanity may appear, varying from a moderate mania to wild and unconscious delirium."

The second manifestation of epileptic disease is in the perversion of the moral faculties. "Epilepsy leads to depravity," says a recent author, and its truthfulness is attested by all having experience with this form of disease. When little or no intellectual aberration is apparent, the moral qualities seem to be blunted. This moral blindness, which contrasts strongly with the individual's previous character, may continue as the constant state, or, which is the more common, may be of variable duration, and of periodical recurrence. It occurs occasionally as the first manifestation of epileptic disease, and may occur at regular intervals for a long period before convulsions make their appearance. The subsequent appearance of the convulsive seizures offers an explanation, if not justification of their conduct. In this class, lying, thieving, and the whole category of minor crimes is indulged in, seemingly out of pure love of crime; little regard is paid to law, either human or divine, and they figure constantly on the police dockets. Maudsly associates moral insanity more frequently with epilepsy than with any other disease. Epileptic criminals prove wholly incorrigible and nothing is more significant than the fact that in spite of all punishment they persevere in their criminal ways. This state of affairs, however, soon changes, the intellect becomes involved and undoubted insanity results.

---

* Op. Cit. ; p. 76.

We now proceed to examine that form in which the excitement extends over a longer period, rarely, however, persisting longer than two or three weeks at a time. Insanity into which epilepsy enters, either as a cause or complication, has, so to speak, a physiognomy of its own. It may manifest itself in any one of the three grand divisions of mental disease, mania, melancholia, and dementia, but it so impresses its peculiar type upon the symptoms, that the epileptic element can be readily diagnosed. Of the mental condition of the asylum epileptic in the interval between his paroxysms of excitement, it is unnecessary to add to what has already been said, except to state that the peculiarities then spoken of are all increased. He is forgetful, irritable, suspicious, fault-finding, liable to outbursts of anger and impulsive actions of all kinds.

The peculiarities of epileptic insanity are, first, the abruptness of the invasion. Often the attack can only be foretold a few minutes previous to its occurrence; the more common prodromes are changes in his disposition or feelings, or pain in his head. The onset may be violent, or the patient may leave off his occupation abruptly and wander here. and there without any definite idea—a vague feeling of restlessness comes over him, a desire to go, whither or why he knows not. "This impulsive want to wander about," says Jules Falret, "is nearly constant in this mental state, and deserves to be carefully pointed out." In this state, he is subject to gloomy fears and forebodings, and often, a word spoken, the presence of certain persons, or it may be a mere subjective sensation, is sufficient to call forth the most murderous attacks. These explosions of violence, during which homicide or the most atrocious crimes may be committed, of which we have already spoken as constituting in some cases the only tangible departure from mental health, are equally characteristic of this form of the disease. They also occur in the demented. E. M., perfectly demented, whose insanity dates back a quarter of a century, has convulsions only at long intervals, but is subject every few days to terrible paroxysms of fury. With a scream she will throw herself on the floor, strike her head, kick and bite, and talk incoherently about graveyards. In a few minutes the paroxysm ends as quickly as it began, and she returns to her normal condition. The attack, though violent, is brief, and generally ends suddenly, though not with the abruptness that marks the onset. The culmination is sometimes reached in an explosive act, after which the patient wakes up as from a dream. In one case, where the excitement had extended over several weeks and who was especially dangerous to his companions, the patient struck a person in no way interfering with him, a violent blow over the eye, and almost immediately the attack ended. Other cases run a different course, and present merely the volubility of words and restlessness of action of ordinary mania, and still rarer cases present the depression of melancholia. However, these cases usually present greater perversion of the moral qualities, are very suspicious, often manifest erotic tendencies, and are more irritable and impulsive in their doings, than similar cases not having the epileptic element.

There are two other peculiarities of these attacks of mania which deserve especial mention from their judicial importance. I refer to the coherency of thought manifested during the period of excitement, and the imperfect recollection afterwards of things said and acts done. The observations of Jules Falret are so pertinent that I quote the following: * " In spite of the disorder and violence of their acts, *their language is in general considerably less incoherent than that of many insane individuals.* It is surprising how easily, in spite of their state of agitation, one can follow the train of ideas expressed by epileptics. Their delirium is more connected and comprehensible than is usual in mania. They

---

* Quoted by Trosseau. Clinical Lectures ; p. 76.

understand better the questions put to them, they answer them more directly, more exactly, and notice what goes on around them more frequently than most insane persons suffering from general delirium with excitement. The less marked incoherence of the delirium and the greater distinctness of ideas during the attacks, are all the more remarkable that they singularly contrast with the nearly total obliteration of all recollection of the fit after it is over, a defect of memory which is also an almost constant symptom of the attacks of epileptic mania."

The following cases illustrate the above: M. S. has been subject to fits for the last eight years, rarely escaping longer than a week at a time. She had several seizures during the day; suddenly she sprang up from the bed, upset everything in her room, and ran up and down the ward, praying in a loud voice for deliverance from the evil-doers around her. It became necessary to remove her to another ward, at which she became very indignant, refused to go, declaring that she was removed because her clothes were not as "costly as some fine ladies," that we had long tried to cheat her out of her rights. She put a false interpretation upon every explanation that we offered, charged the nurses with every kind of misdeed, and became very obscene in her talk. Yet her statements were so coherent and her accusations so plausibly supported, that it seemed like nothing but an exhibition of bad temper. In a few days the excitement passed off; she was surprised to find herself in another ward, and had no remembrance of her removal, or of her sayings at the time, and felt grieved when told of them.

. A. B., æt. 26, has been subject to epileptic vertigo and night attacks of epilepsy since her twentieth year. Several times her shoulder has been dislocated during the fit by the violence of the muscular contractions. She had shown no symptoms of mental derangement, aside from a failure of memory and occasional despondency, which was the more noticeable from her naturally cheerful disposition. One Sunday morning, a few months ago, she went to the barn, seized an axe that stood near, and placing her head against a wagon wheel, struck several blows on her forehead and head, making some ugly wounds. She fell senseless, was soon discovered by her friends and carried into the house. She remained in this condition until her wounds were dressed by the surgeon, when she awoke and became so violently maniacal that it was necessary to confine her hands to prevent her from tearing off the dressings and injuring others. The excitement passed off in a few hours, and she had no recollection of her suicidal attempt or her after acts, and seemed much astonished when told of them. She was admitted into the asylum before her wounds were healed, and has since shown no evidences of insanity farther than the failure in memory and the irritability peculiar to the class.

There are three points of special interest in the case: 1st, She had no fit for several days either preceding or succeeding the suicidal attempt, which in all probability took the place of the regular convulsion. 2d, The interest which attaches to it in a medico-legal view. Had the impulse been homicidal instead of suicidal, how difficult it would have been to convince the court with the present ruling that she was an irresponsible agent. 3d, The fact that she has never experienced but the one paroxysm.

Morel first described a form of epileptic insanity to which he gave the name of *epilepsic larvee.* We have already alluded to it incidentally, but its importance in a legal point of view entitles it to a more careful consideration. In this form, the common convulsive phenomena are absent, but the mental phenomena are so characteristic that the class to which the disease belongs cannot remain in doubt. Morel[*] gives as the peculiar symptoms: "Periodical excitement followed by prostration and stupor; excessive irascibility with-

---

[*] Jour. Ment. Science. January, 1863.

out cause; the manifestation of aggressive violence, marked by instantaneity and irresistible impulse; exaltation of the sensibility; homicidal and suicidal tendencies; intercurrent insane ideas connected with the state of cerebral excitement; exaggerated notions of physical power, of wealth, of beauty, or of intelligence; erotic tendencies coupled with exalted religious feeling; hallucinations of terror; sensation of luminous atmosphere; horrible dreams or nightmare; gradual progressive debility of the powers of understanding, especially of the memory; loss of recollection of events transpiring during the paroxysms, the insane symptoms of each periodic attack having, both with reference to the ideas which occupy the mind and to the actions committed, the same identical character; and lastly the violence and duration of the delirious excitement determined by the duration of the remission." Some observers, among whom is Dr. Sankey, refuse to recognize the disease under consideration as a distinct species of epileptic insanity, but rather consider the cases adduced in its support, as examples of periodic or recurrent mania. The natural history of the two forms, however, presents many points of contrast. The epileptic form is marked especially by the abruptness of its invasion, the uncontrollable impulses to violence, oftentimes homicide and suicide, the brevity of the attacks, the absolute resemblance in deed and act of successive attacks, and the suddenness of the intermission. In ordinary recurrent mania there is a better marked period of incubation, during which the maniacal symptoms gradually appear, the attack gradually increases in severity, rarely stopping short of weeks, it may continue for months without abatement. Its force expended the patient is left for a variable period afterward in a depressed condition, from which he gradually passes into his normal condition. Though recurrent disease is not infrequently associated with acts of violence, they do not possess the furious and vicious nature of epileptic impulses. Very often the diagnosis is confirmed if the earlier history of the patient can be traced and we find him to have formerly suffered from epileptic seizures, or again, if convulsions should appear in the subsequent course of the disease.

I present the following cases:

H. K., æt. 18, was admitted in an attack of mania. The accompanying medical certificate said he was not subject to epilepsy. For several months he had at times shown symptoms of mental derangement; he had been inclined to wander about, quarreled with persons around him, became irritable and easily excited. For ten days previous to his admission he had been very violent. Ordinarily he presented the restlessness of action and thought peculiar to mania, but at times he was seized with a perfect fury, would destroy his clothing, break up the furniture of the ward, and spring furiously upon any one who chanced to be near him, using any weapon at hand. He gave no premonitions of his attacks. He was very revengeful, and after the beginning of the attacks would seek out those whom he disliked and attempt to injure them. On removing him to his room, which was not often done without difficulty, the excitement soon passed off. He remained in this condititon about six weeks, when, after severely cutting the face of an unoffending patient, he seemed to wake up and speedily recovered. He had a very imperfect recollection of coming to the Asylum, or of his doings while here. In two weeks afterward, he had a well defined epileptic convulsion, and was subject to them at irregular intervals during the remainder of his stay in the Asylum. In tracing back his history it was ascertained that he had at least two epileptic seizures about a year before. During the eight months in which he exhibited irregularities of conduct and had an attack of mania, the seizures did not recur, but made their appearance immediately after the mental improvement.

M. G., æt. 23, when admitted had never experienced a

regular convulsion, but had for some time past been subject to occasional attacks of "dizziness or blindness." Her disease is paroxysmal; during the intermissions she is usually pleasant and assists in the work of the ward. At times she will jump up, run down the ward, break a window, and then quietly resume her seat. Again she will suddenly leave off whatever she may be doing, become abusive to the attendants whom she has just been assisting, strike persons, saying they are making faces at her, declares she is going off, and makes every attempt to get out. Several times she has succeeded, in spite of efforts to prevent her, and on one occasion she ran quite a distance over the fields with barely enough clothing to cover her nakedness. She kicks, bites and fights terrifically when returned. Usually she becomes stupid and drowsy, and rapidly comes out of her paroxysms with a very indistinct recollection of her doings. In her better periods she says she does not know what makes her act so, that she cannot help it, and does not want to go home until she is well. Her paroxysms recur frequently, and all bear a remarkable resemblance to each other. During the first remission she attended a dance in the chapel. She seemed to be highly pleased, when without a single premonition she gave a scream, and struck the patient sitting next to her. On being removed from the chapel, she escaped from the nurses, and getting into a dining-room where the table was set, assaulted them with dishes, and was soon master of the situation. The last two paroxysms passed off with tolerably well marked convulsions, and the drowsiness of previous attacks.

Dr. Thorne Thorne, in St. Bartholomew Hospital Report, 1870,[*] relates the following case, which I abbreviate: "H. S., æt. 36., was admitted for an attack of bronchitis. He seemed to be strange in manner shortly after admission, and on several occasions said his wife was looking in at the ward

---

[*] Quoted in Jour. Ment. Science; January, 1871.

window. One night he suddenly jumped out of bed, and rushed wildly to the door, which he had no sooner opened than he fell prostrate on his back. He seemed to retain consciousness, but offered no explanation of his conduct.

Occasional attacks of mental excitement similar to the above, occurred during his stay in the hospital. It was ascertained that for several years he had complained of severe pain in the head, and of dimness of sight and trembling. Four or five years before admission, he had several well-marked epileptic seizures, which have constantly recurred, and now average about every three weeks in frequency. Prior to his entering the hospital he had never shown any morbid mental symptoms. Several times since leaving the hospital he has suffered from mental depression which passed into maniacal excitement. His countenance becomes wild, intellect confused, and he will snatch a knife and threaten to kill his children. He rushes after them as they in terror seek to hide themselves. Several times when his wife had locked herself in the room with him, it has required all her strength and tact to prevent him from throwing himself out of the window. This excitement may last from several hours to an entire night, when he gradually becomes quiet, and usually sleeps. On awaking, he has but a dim recollection that he has been ailing, and none whatever as to what has been said or done. The patient often returned home without being able to give an account of his doings, and in these states has been guilty of petty thefts. At one time, while in this stupid state, parcels of scented powders were found on his person, which he could have had no object in purchasing, and of which he emphatically declared he knew nothing. Dr. Thorne very properly remarks, "it is impossible to overestimate the importance of such a case from a medico-legal point of view, because, though persons suffering from masked epilepsy may be as this man was, and still is, able to follow their daily avocations, yet they must necessarily be considered as placed, at least during the continuance of their attacks, beyond the category of healthy and responsible minds. Volition is in abeyance, and hence responsibility must be also."

The *prognosis* of epileptic insanity as to permanent recovery is exceedingly unfavorable. So far as recovery from the particular maniacal excitement is concerned, the prognosis is good, but with the presence of the same cause, and that cause a progressive one, the liability to relapse is very great.

Several cures are recorded by good observers, but these must be received with some degree of allowance, since the mental and physical phenomena of epilepsy may become latent for a considerable length of time, inducing false hopes in its victims, and then break out with accumulated severity. It is very doubtful whether the mind ever thoroughly regains its tone after it has once become affected through the seizures, yet the periods of remission from active symptoms do occasionally extend over months, and even, in some recorded cases, for years.

The prognosis as to life is unfavorable, the larger number dying before the middle period of life is reached.

Last year I reported to this Society on " *The Temperature of Certain Nervous Diseases,*" and gave a table of thermometrical observations on insane female epileptics. Further experience in the use of the thermometer has confirmed the general conclusions then arrived at, and which may be briefly summed up:

I. The temperature of epileptic insanity is high, the evening altitude being in excess of that of the morning.

II. The temperature of the same individual varies largely and bears a relation to her mental condition at the time the observation is made. The periods of excitement present an increase, often amounting to several degrees over the usual temperature.

III. A depression of temperature occurs just before a

_____, ich is better marked in the sleep that follows the seizure; several hours usually elapse before the inter-paroxysmal range is reached.

- IV. A succession of fits increases the temperature, and several days may elapse before the thermal equilibrium is restored.

We do not propose a thorough discussion of the jurisprudence of insanity, even when complicated with epilepsy, this is already in the hands of a special committee; but merely propose to point out the medico-legal aspects of one factor, epilepsy.

The courts have for a long time ruled the irresponsibility of acts committed in close connection with the convulsion, and tacit jurisprudence has limited the time to four days — two days preceding, and two succeeding the fit. This ruling we hope to show is wholly at variance with our present knowledge of the mental condition of epileptics, and with the ordinary psychical phenomena of the disease. We have already shown that the sanity of epileptics at any time after the development of the disease is an open question. Beyond doubt the character of the person so afflicted undergoes modification. This change can sometimes be noticed after the first seizure, becoming better marked as they recur, and is evidenced in both their mental and moral natures, the latter not infrequently betraying the greater alteration.

Dr. Ray * writes, "the fact of its (epilepsy) existence being established, is it going too far to say that legal responsibility is presumptively annulled, and that the burden of proof lies on the party that alleges the contrary? People are scarcely prepared for it yet, perhaps, but to that complexion they will come at last."

All epilepsy, according to Marc, warrants upon the event of a criminal act the suspicion of mental disorder, which suspicion is increased in the absence of personal motive.

* American Journal of Insanity. October, 1867.

Dr. Bucknill, in the preface to his Prize Essay, * says: "The existence of habitual epilepsy might justify the decision that a violent and unprovoked assault was not the result of criminal passion, but of uncontrollable excitement of the brain. * * * The psychical phenomena of epileptics are of a kind to set the metaphysicians at defiance; simple, uncontrollable anger, with or without some slight cause, and accompanied by the most reckless violence, or gloomy and cruel revenge for some supposed injury, or simple, motiveless rage, all without delusion or the presence of any of the metaphysical tests, and quickly passing into an ordinary and apparently a healthy state of the mental functions are the psychical characteristics of this fearful and obscure disease."

Dr. Burgess † holds that "a person may act and conduct himself rightly for a long period; in epilepsy he may do so for months, and even years, and although quick to evade his particular hallucination, and point out and oppose the errors of others, he may be suddenly attacked with mania himself, and neither the epileptic nor those afflicted with intermittent madness can at any time be held responsible."

Baillarger concludes a lecture before the Academy of Medicine (Paris,) with the following propositions: ‡ First, That besides declared insanity, there exists with certain epileptics a special moral and intellectual condition. Second,. The medical jurist ought in many cases to apply himself to make apparent the principal traits which characterize this condition, to extenuate, at least as much as possible, the responsibility of the sufferer."

Wharton and Stille § hold, "Persons truly epileptic are easily excited to anger and revenge on the slightest provocation, in the intervals between their attacks. Although their

---

* On Criminal Lunacy, p. 37.
† Op. cit., p. 104.
‡ Med. Critic and Psych. Journal. Vol. I, p. 512.
§ Medical Jurisprudence. P. 117.

attacks do not always attain to such a degree as to deserve
the name of mental derangement, yet it should never be for-
gotten that there is always a morbid predisposition to insane
ebullitions, and in general a morbid irritability which must
impair, if not destroy, the moral responsibility of actions
growing out of them."

In opposition to the above views, we have Caspar* ask-
ing, "How completely objectionable to declare epilepsy to
be a disease which renders those affected with it irresponsible agents, as is done by the earlier authors and their modern compilers—the fact taught by experience that epilepsy
occurs in nature in innumerable degrees of intensity, is op-
posed to the general applicability of any such dogma."

We do not believe that all the acts of epileptics should be
excused because of the presence of epilepsy, as will appear
further on.   Dr. Caspar founds his belief upon the assertion
that the mania of an epileptic patient does not differ psycho-
logically from ordinary mania, whereas we have endeavored,
successfully or not, to show that epileptic disease does manifest itself in peculiar psychological phenomena, and differs
in this very respect from mental irregularity dependent
upon ordinary causes.

Dr. Echeverria † writes: "Believing that there is a morbid obliteration of moral feeling in epileptics and that they
are mastered by uncontrollable impulses, I would hesitate to
fix the limits of their responsibility. Chronicity, to my
mind, is as essential as any of the other characteristics of
epilepsy, wherefore I would be far from thinking a person
epileptic because he or she were troubled at some previous
time with epileptiform convulsions that have not reappeared.
Consequently, I reject exculpating any overt act attributed
to epilepsy unless committed in clear relation with the spasm
or while the perpetrator exhibited plain signs of cerebral

---

* Forensic Medicine.   P. 188, vol. IV.
† Op. cit., p. 362.

epilepsy; and whichever be the medico-legal faith given to this last form of the disease, it is nevertheless positive that epileptics undergo a moral degeneracy through their malady and that we are never safe with them, as has recently been asserted with great truth by Delasiavue." He gives further on in the chapter two cases. I abbreviate the following: "One morning an epileptic made an assault on one of his companions with a knife, for which no provocation had been offered. He was subject to nocturnal attacks, but no mention is made of a fit having occurred the previous night. The mental trouble in this instance is not apparent to any person who would not inquire closely into the patient's symptoms."

Our courts have occasionally recognized the sufficiency of epilepsy as a defence in capital crimes. In the trial of Fyler for the murder of his wife, in Onondaga county, New York, 1855, the defense of epilepsy was entered for the first time in this country. Fyler had for several years been subject to epileptic convulsions, but without producing any obvious mental impairment beyond the immediate effect of the fit. No evidence was offered to show that a fit had occurred about the time of the homicide, or that he exhibited any of the ordinary signs of a fit. The trial, after various delays, came on a year after the commission of the crime. He was found guilty, but a committee appointed by the court to investigate his mental condition, adjudged him then insane, and he was remanded to an asylum.

Dr. Gray, Superintendent of the New York State Asylum, relates the following case: * An epileptic was on trial for the murder of his wife. The plea of insanity was entered, but the counsel having given their entire attention to the fact of insanity and little or none to the question of epilepsy, the evidence did not warrant the experts in considering the case one of insanity or irresponsibility. The pris-

* Amer. Jour. Insanity. October, 1870.

oner had a well-marked epileptic seizure in court. Dr. Gray, who sat beside him, wrote a note to the judge, stating that the prisoner was in an epileptic state, and unconscious of what was going on. The jury returned a verdict of guilty, but the court did not pronounce sentence. Under a subsequent investigation, the prisoner was committed to an asylum, on the ground that he had been an epileptic, was still an epileptic, and was, therefore, possibly a person of doubtful responsibility.

Again, the convulsions may occur at night, and the prisoner or his friends be unaware of the fact that he is an epileptic. Baillarger gives the case of Councillor Lemke as an instance of epileptic furor. It was this case that gave rise to the division of mental disease known as Mania Transitoria. The Councillor was seized with a sudden fury to kill in the middle of the night. He attempted to throw his wife out of the window, when assistance coming, she was released. His stertorous breathing a few moments before had awakened his wife, and her efforts to relieve him called forth the furious attack.

Trousseau was consulted by a newly-married couple under the following circumstances : The wife, soon after their marriage, was awakened by the strange movements of her husband. Suddenly she was attacked, and had her screams not brought help, she would have been severely injured. On a subsequent occasion, she was awakened in time to strike a light, witness the convulsion, and escape from the fury which immediately followed. Epileptic attacks were, up to this time, unsuspected.

Dr. Delasiauve * relates the case of a man who was once a patient in the Bicetre, and was subsequently returned after killing his wife. His occasional aberration of mind was plead in his defense and prevailed. It was only in the asylum on his second admission, where, save a few transient

---

* Noted in Amer. Jour. Ins.   October, 1855.

agitations, he exhibited perfect lucidity, even to his death, occurring a long time afterwards, that his momentary wanderings, and, in part, his criminal acts, were traced to night attacks of epilepsy.

Irresponsibility should not be made dependent upon the convulsion, since *vertigo* or *petit mal* may be followed by as impulsive actions and as disastrous results as the general convulsion. Mare, with many other authors, places prominently among the conditions of immunity, the intensity of the paroxysm. Cases, however, are not infrequent in which there is no convulsion, and even the loss of consciousness is unknown to the person himself, and only observable by others in the sudden stoppage in the midst of a sentence or an act. Yet in this unconscious interval lewd words may be spoken, laughter indulged in, incongruous or criminal deeds performed, and no recollection of them remain. Certainly the unconscious criminal should not be held responsible, though he seems sane a moment before and after.

As examples, I mention an epileptic architect who does not hesitate to ascend the loftiest scaffolding. When he has a fit, he runs rapidly over the scaffolding, shrieks out his own name in a loud voice; in a quarter of a minute he resumes his occupation and gives his orders to the workmen. Unless he is told, he does not recall his singular act.

The President of a provincial tribunal, while holding court, abruptly left his seat, muttering a few unintelligible words, and shortly returned. He had no recollection of having left his seat. Going out in a similar way on another occasion, the usher followed him. He went to the council room, made water in the corner, and returned, perfectly ignorant of his incongruous act.*

An out patient of the London Hospital has in connection with his epilepsy subjective sensations of smell. He has had convulsions in which he has bitten his tongue, but re-

---

* Trousseau. Op. cit.

cently there have been losses of consciousness without movement of the limbs. In these, he is unconscious, becomes pale, and usually remains still. In one attack, he went from Poplar to Deptford—eight miles. The facts are, he lost himself at Poplar, and found himself again at Deptford where, fourteen years ago, he used to live. Sometime since he walked into a canal during a seizure.*

I was recently consulted in the case of a lady, Mrs. B., æt. 50, who for a year past has been subject to what her friends call "fits of absence." She often stops in the middle of a sentence or an act, stares fixedly, and not infrequently makes remarks having no connection to the previous conversation, or hides away articles near at hand; in a few seconds she is able to finish the sentence or proceed with her work wholly unconscious of the interruption. These seizures occur at irregular intervals, but are gradually becoming more frequent, and have lately been attended with more or less twitching of the facial muscles. To give an instance: At the close of services on the previous Sunday, she experienced a seizure just as the parson's wife, Mrs. A., spoke to her. Mrs. B. shook hands, and remarked, "By high heaven, it shall not be done, I warn you." She passed half the length of the church aisle before she recovered. Then, after saluting some friends, she turned again to Mrs. A., shook hands cordially, expressed pleasure at meeting her, and engaged in conversation, without knowledge of her actions but a few minutes before. Aside from a greater irritability and suspiciousness, her friends, as yet, can detect no mental irregularity, except such as is connected with her seizures. I could adduce several cases confirmatory of the above proposition, coming under my notice in the Asylum, but prefer to offer the above examples, because the sanity of the parties was not questioned.

Dr. Pritchard and others hold that freedom from epileptic

---

* London Lancet, March 18, 1871.

symptoms for a considerable period renders the individual
amenable to the law. We have endeavored to show that
one of the chief characteristics of epileptic disease is its
periodical and sudden recurrence; the remissions may be
well marked and may continue for a long period, and yet
all of the phenomena may reappear with accumulated severity on the slightest cause. Dr. Burgess* says upon this
point: "Insanity being in the system, as in *epilepsy*, either
fully developed or in a state of incubation, the paroxysmal
madness is liable to commence or to return at any moment,
and it is difficult to say when a lucid interval exists, when
it is gone, and when it will return."

Dr. Burrows † relates the case of a quiet, sober and industrious tradesman, subject to fits of epilepsy. He was sitting
calmly reading his Bible, when a female neighbor came for
some milk. He looked wildly at her, seized a knife, and attacked her, and then his wife and daughter. Their cries
brought assistance, and he was secured before he had inflicted
any fatal wound. After the paroxysm ended he had no recollection of his acts. A year before, he had experienced a
slight attack; nine years have since elapsed, without a recurrence of epilepsy, or disturbance of his mental faculties.

The strongest reason for the non-association of the crime
and fit in casting up the prisoner's responsibility, is the frequency with which the furor or madness replaces the fit,
constituting the different grades of masked epilepsy. We
have already referred to this in different connections, and it
is unnecessary to multiply examples; suffice it to say, that
no fact in psychological medicine is better established than
that the mental phenomena do, at times, replace the convulsions. The attack is characterized in most cases by unconsciousness, in others, by an entire forgetfulness of the act
afterwards, yet these are not essential, as in the case of the

---

* Op. cit., p. 105.
† On Insanity, p. 156.

epileptic who in one paroxysm murdered several persons successively as he met them on the road, and had afterwards a perfect remembrance of each murder.

Now since the existence of these cases is undoubted, why should the law not recognize them? Because they seem sane just before and after the commission of the act, and are then abundantly able to judge between right and wrong, proves absolutely nothing. The ungovernable impulses are the outgrowth of a morbid condition of the nerve element very like the fit, and the convulsive impulses of the brain are as much removed from the control of the individual as are the convulsive movements of the limbs. It would certainly be as reasonable to hang a man for not keeping his limbs quiet, as to hang him for not keeping his brain quiet. The epileptic furor is not reasoning, it does not, as a rule, recognize right or wrong during the paroxysm: there is little room for doubt but that the criterion of responsibility set up by the courts is founded in error and results in great injustice to this unfortunate class.

Dr. H. Tuke, at a meeting of English lawyers in 1865, called to consider the justice of capital punishment, said that fifty-four asylum physicians had combined to combat the doctrine that the legal responsibility of the insane depends upon their knowledge of right and wrong. They agreed that "the belief in the responsibility of the insane based on their appreciation of the right and wrong, is irreconcilable with facts known to all alienists, because it is clear that this appreciation often exists in individuals whose insanity is undoubted, and it is even observed together with insane ideas and those of a dangerous and impulsive character." The German and French Alienists have almost universally expressed a similar opinion.

It is not alone the ability to decide between the right and wrong, but the power to do the right and abjure the wrong, that should determine one's responsibility. Inasmuch as the cases are few in which the courts have ruled the irresponsibility of epileptics in the absence of the fit, those cases, of which many are recorded, in which suicidal and homicidal impulses have been felt, but through physical restraint or by the exercise of great will power have been overcome, are deserving of careful study. They are entitled to greater credence, because the same urgency for deception does not exist as when the individual is arraigned as a criminal before the judiciary.

Esquirol quotes the following: A peasant, æt. 27, had been subject to epilepsy since his eighth year. For two years the disease changed, became masked, and instead of the convulsions from which he had previously suffered, he was periodically seized with an almost irresistible desire to murder. He could foretell a paroxysm for hours, and sometimes for days before its invasion. From the moment of this presentment, he earnestly demanded to be tied or chained, in order to be prevented from the commission of crime, the moral nature of which he perfectly appreciated. He would suddenly shriek out, "Save yourself, mother, or I shall choke you." When bound, he made frightful grimaces, sang, &c. The paroxysm usually subsided in a day or two, when he would ask to be released and be thankful that he had injured no one.

Dr. Echeverria * relates the following case: A woman subject to grand and petit mal, has slight attacks of vertigo during the day, preceded by a vision of a sudden dash of fire. Her memory has failed, and she complains of feeling insane without a will. She has at times blind impulses to strike her child, and being unable to resist them, her mother has been obliged to take the child away from her.

That form of disease known as Mania Transitoria, and which has been the cause of so much bitter discussion among alienists and jurists, is probably in a large number of cases

---

* Op. cit., p. 370.

a manifestation of epileptic disease. If the history of those cases usually quoted as examples of transitory mania, be traced, quite a number will be found to present evidence of epilepsy. The modern view that muscular contraction is not essential to epilepsy, but that a momentary loss of consciousness may be the only sign of an attack, enables us to better appreciate these cases. We have already shown that the mental phenomena are not in proportion to the violence of the muscular seizures, but on the contrary may often bear the very opposite relation to each other. We do not affirm that all cases of transitory fury are epileptic. It must of course be borne in mind that all forms of nervous disease are more or less paroxysmal, and that hereditary insanity in particular is especially liable to explosive actions similar to those that characterize epileptic insanity.

The increased susceptibility by which internal and external impressions become altered and transformed in the passage to the sensorium deserves notice. In the peculiar temperament which epilepsy begets, an act, word, excess in living, or any irregularity which would scarcely disturb the equilibrium of a healthy organism, taken in connection with the terrible hallucinations and illusions which the epileptic experiences, and which the diseased sensory ganglia may connect with surrounding persons and things, are sufficient to instigate him to the most cruel and violent acts. If nothing more, the knowledge of this fact should extenuate the punishment meted out to his crimes.

These facts almost lead us to adopt the sweeping conclusion of Trousseau: That when a man commits a murder on a sudden impulse and without motive, if that man has not previously shown symptoms of madness, and if he were not in a state of intoxication, his action should almost always be explained by the existence of epilepsy.

Dr. Maudsley* says: "Whenever a murder has been

---

* Body and Mind, p. 72.

committed suddenly, without premeditation, without malice, without motive, openly and in a way quite different from the way in which murders are commonly done, we ought to look carefully for evidence of previous epilepsy, and should there have been no epileptic fits, for evidence of an aura epileptica and other symptoms allied to epilepsy."

Epileptics, in addition to their pathological infirmities, are subject to all the other infirmities of our common humanity, and hence may be guilty of misdemeanors and crimes independently of their disease. Whether any special misdeed or exhibition of anger is the result of criminal passion, or the outgrowth of disease, bearing constantly in mind the liability of these paroxysms to attend epilepsy, must be decided upon the data for deciding upon insane acts in general : i. e., the relation the parties sustained to each other ; the character of the crime ; the manner and place and time of its execution ; the conduct just before and after its commission, &c.

When we remember that epilepsy, if not leading to undoubted insanity, brings about, in its very inception, a change in the social, moral and intellectual life of the individual ; that the will power is sapped and the individual is to a greater or less degree governed by morbid impulses : when we recall the sudden and transient nature of the maniacal attacks, and the liability of homicidal acts to attend them ; especially when we consider that the criminal act may replace the convulsion, and that apparent sanity may exist just before and after its commission, we are impressed with the necessity of great care in passing upon the legal responsibilities of this class of society.

Jurisprudence has, however, a two-fold object : the protection of society as well as the protection of the criminal.

The existence of epilepsy established, it is probably going too far to excuse the party from all responsibility, but it certainly should be held to have this effect until the contrary is proven, until it be shown that the criminal act was not the offspring of the disease. Society rightfully asks that it be protected from a recurrence of the crime. Humanity, with an equal right, asks that if the crime is the result of disease the prisoner shall be spared the disgrace of punishment. An asylum is, beyond question, the proper place for the large majority of epileptic criminals. We would also recommend this disposition in those doubtful cases which so perplex expert and judge, since in these cases, a careful watch, oftentimes for a considerable period, is the only way to mete out justice. If it should appear that the disease is not father of the crime, the ends of justice would not suffer by the delay. It is not the swiftness of punishment, which often savors of revenge, but the certainty with which the offender receives his just rewards, that deters criminals and upholds the majesty of the law.

In conclusion, permit me to urge upon this Society, so noted for its appreciation of all public charities, the claims which this unfortunate class have upon us, and to ask you to second the recommendation of the Board of State Charities for the establishment of a special hospital for epileptics. The necessity of such an institution is seen in the present deplorable condition of the epileptics in the jails and infirmaries of the several counties. The number of this class at present in asylums, jails and infirmaries is about five hundred. Of this number, three hundred and fifty are in the jails and infirmaries, of whom, according to the last annual report of the Board of State Charities, fourteen are in close confinement, two are chained, and eight are quite filthy. I have no means of estimating the number of epileptics in their homes throughout the State, but from the frequent application made here for the admission of insane epileptics, I infer the number is not inconsiderable.

Another object gained by such an institution would be the relief afforded to our overcrowded insane hospitals. Even when the asylums now in process of enlargement and con-

struction are ready for occupancy, they will be insufficient to
accommodate the insane, exclusive of the epileptic insane.

Every one will readily recognize the impropriety of re-
ceiving epileptics into the insane asylums: First, Because
the restlessness of thought and action in mania proves a con-
tinual source of irritation to morose epileptics. Second, Because of the deleterious effect upon an already disturbed
mind, or one just struggling back to health of witnessing
the frightful appearance and apparent suffering of one convulsed.

In an institution designed expressly for this class, some
modifications could be made in the present plan of hospital
construction, so as to better meet their special wants. The
suddenness of an epileptic seizure and the fall which attends
it, render some provision necessary against accidental injury
to themselves. Still more, epileptics, except in very rare
cases, should sleep alone. The associated dormitories so
common and so commendable in our insane asylums, are un-
fitted for epileptics.

No fitter commentary can be made than to cite you to the
sad case that occurred in the Lawrence County Infirmary.*
Two epileptics, both subject to night attacks, were crowded into the same room and bed. "One morning the young-
er of the two was found strangled to death." The old man
is now serving his life sentence in the penitentiary.

* Rep. Board of Charities. 1869.